Contents

INTRODUCTION 4

ASTRONOMY FOR CHILDREN 6

AMATEUR ASTRONOMY FOR KIDS WITH BEGINNER TELESCOPES 10

HALLOWEEN COSTUMES FOR KIDS AND INFANTS 13

HALLOWEEN COSTUMES FOR KIDS 16

CHOOSING THE BEST CHILDREN'S TELESCOPE - TIPS FOR BEGINNER TELESCOPE BUYERS 20

SPACE THEMED BIRTHDAY PARTY IDEAS FOR KIDS 22

DESCRIPTION OF THE ASTRONAUT JOB 25

LET AN ASTRONAUT TELL YOUR CHILDREN A FAIRY TALE 28

UNCONVENTIONAL WISDOM FOR ASPIRING ASTRONAUTS 34

THE ASTRONAUTS WIVES CLUB - A TRUE STORY 37

HOW TO GET BIGGER BY IMITATING ASTRONAUTS 41

RISKS FOR ASTRONAUTS 44

LIVE YOUR ASTRONAUT DREAMS AT THE KENNEDY SPACE CENTER 47

DISCOVER DISTANT PLANETS WITH YOUR ASTRONAUT COSTUME 51

THE ROBOTIC ASTRONAUT ASSISTANTS 54

ASTRONAUT COSTUMES ARE A FUN WAY TO PAY HOMAGE TO OUR HALLOWEEN STORY 57

DISCOVER NEW THINGS IN AEROSPACE TECHNOLOGY 60

"ESCAPE FROM PLANET EARTH": FUN FOR CHILDREN AND ADULTS 64

CHILDREN INTERESTED IN ASTRONOMY - SPACE GAMES FOR CHILDREN 69

CONCLUSION 75

INTRODUCTION

All mankind has spent nearly more than 2000 years understanding the universe and the solar system in which we live. Researchers have recorded what they have learned about the universe for thousands of years. The curiosity that some astronomers know about the universe has revealed many of the secrets of the sky. Ancient man always wanted to know something about the universe and explore space. In 1959, NASA (National Aeronautics and Space Administration) selected seven people in the United States to be the first astronauts to travel in space.

The first project in human history that allowed humans to orbit around the Earth was the Gemini and Mercury project, whose main task was to put a spaceship into orbit, stay there and safely with the spaceship Return the surface of the earth. The famous Apollo program was launched in 1961 with the mission of landing on the lunar surface. It took several years for research and test flights when astronauts Neil Armstrong, Aldrin and Michael Collins finally landed on the lunar surface in

1969. The flight that successfully landed on the moon was Apollo 11.

After the successful landing of Apollo 11, several missions with lunar flight programs such as Apollo-Soyaz, Space Shuttle and Skylab were made on the moon. These missions are supposed to land on the moon, but we never landed on another planet or object such as the asteroid. NASA has sent a series of unmanned spacecraft into the solar system to gather information as well as about planets and Mars. There is a long list of unmanned spacecraft such as Viking 1 and 2, Mariner, Pioneer 10 and 11, Explorer, Magellan, Venus 1 and 2, Voyager 1 and 2, Galileo, Lunar Orbiter, Clementine and Surveyor.

ASTRONOMY FOR CHILDREN

What Is An Astronaut?

An astronaut is an individual who is uncommonly prepared to go in space. Astronaut onboard a rocket can perform various undertakings. For the most part, there is an authority which drives the strategic a pilot. Different positions may incorporate flight engineer, payload authority, crucial and logical pilot. Astronauts must undergo intense training and testing before they can take part in a space flight. They must demonstrate that they are able to manage the physical tension from the high gravity of the launch to the zero-gravity orbit. They must also be tech-savvy and capable of dealing with stressful situations that may arise during the mission.

Space Suits

Astronauts have special equipment, called spacesuits, which they use when they have to leave the safety of their spacecraft. These spacesuits provide air, protect them from the extreme temperatures of space and protect them from the sun's radiation. Sometimes the spacesuits are tied to the spaceship so that the astronaut does not swim away. In other cases, the spacesuit is equipped with small rocket engines so that the astronaut can navigate the spacecraft.

Famous Astronauts

• Buzz Aldrin (1930) - Buzz Aldrin was the subsequent individual to stroll on the moon. He was the pilot of the lunar module on Apollo 11.

• Neil Armstrong (1930-2012) - Neil Armstrong was the first to stroll on the moon. At the point when he stepped on the moon, he offered the celebrated expression: "This is a little advance for people, a major jump for mankind." Neil was likewise part of the Gemini VIII strategic, two vehicles were effectively docked in space just because.

• Guion Bluford (1942) - Guion Bluford was the principal African American in space. Guion flew four distinctive space transport missions, starting as a mission specialist on the Challenger in 1983. He was also a pilot in the United States. Air Force, where it flew 144 missions during the Vietnam War.

• Yuri Gagarin (1934-1968) - Yuri Gagarin was a Russian cosmonaut. He was the first person to travel in space and orbit the earth. He was aboard the Vostok spacecraft when he successfully orbited the Earth in 1961.

• Gus Grissom (1926-1967) - Gus Grissom was the second American to travel aboard Liberty Bell 7. He was also the commander of the Gemini II, who orbited three times around the earth. Gus kicked the bucket in a fire during a pre-flight test for the Apollo 1 crucial.

• John Glenn (1921-2016) - In 1962, John Glenn was the principal American space explorer to circle the Earth. He was the third American in space. In 1998 Glenn voyaged again with the Space Shuttle Discovery. At 77 years old, he was the most seasoned man to fly into space.

• Mae Jemison (1956) - Mae Jemison was the primary dark space explorer to make a trip to space in 1992 with the Space Shuttle Endeavor.

- Sally Ride (1951-2012) - Sally Ride was the main American in space. She was additionally the most youthful American space explorer to go in space.
- Alan Shepard (1923-1998) - In 1961, Alan Shepard was the second and first American to go in space on Freedom 7. A couple of years after the fact he told Apollo 14. He arrived on the moon and turned into the fifth individual to stroll on the moon.
- Valentina Tereshkova (1947) - Valentina was a Russian cosmonaut who, in 1963, was the primary lady to go onboard Vostok 6.

Fascinating realities about space explorers
- "space explorer" originates from the Greek words "Astron nautes", which signifies "star mariner".
- An estimated 600 million people saw Neil Armstrong and Buzz Aldrin walking on the moon on TV.
- Astronaut John Glenn became an American citizen. Ohio Senator, where he served from 1974 to 1999.
- Alan Shepard has become famous for hitting a golf ball on the moon.

AMATEUR ASTRONOMY FOR KIDS WITH BEGINNER TELESCOPES

Astronomy for kids can be rewarding, educational and fun. With astronomy software, children can take a virtual tour of our galaxy from their computer. For the astronomer, there are many beginner telescopes. From the simple mirror telescope to the motorized refractor for equatorial mounting to the computer-controlled GOTO telescope, which automatically points to every object in the night sky simply by entering information in the laptop keyboard.

Each parent's job is to encourage children to want to learn. Who among us can look at the sky full of stars with amazement and cannot be full of questions? The school doesn't have to be boring if the whole universe can be a classroom. Use your children's interest to observe the stars and turn it into a permanent educational offer. Who knows where it can lead. According to what you know, your son may have the next Galileo or astronaut who will colonize Mars.

But before we get it, let's discuss how you can promote your budding astronomer. A simple image search in any search engine provides breathtaking images of space objects such as nebulae, galaxies and star clusters. Showing kids what's out there can inspire their imagination. The courtyard astronomers, who usually use telescopes, do not create fantastic images as reported by the Hubble telescope. Still, more than satisfy their desire to take images that are worth saving. Children can take advantage of years of experience from other club members and are all ready to share them with beginners. Ask when their next star party is coming and plan to attend. This is an excellent opportunity for the whole family to get involved in the new hobby.

At the star festival, you can see several telescopes in action. If you haven't purchased your kids' telescope yet, you can see it in action here. Each club member has installed their telescopes and takes advantage of it to demonstrate them. This way, you can compare and find the one that suits your child. Club members may have used the telescopes they want to part with at the right price.

The Internet is also an infinite source of information on astronomy for children. Astronomy projects, puzzles, games and even videos will keep all the kids busy for hours and get to know the Milky Way, which we call home. If you read the reviews of people who use telescopes, you can focus on child-friendly telescopes for beginners and find the right model for your weapons astronomer. Happy look at the star!

HALLOWEEN COSTUMES FOR KIDS AND INFANTS

When it comes to kids, ideas for Halloween costumes can't be missing. Children's costumes are often divided into infant and toddler costumes and children's costumes. Getting these costumes in a local store is as easy as putting them online. There is a lot to choose from, and you are sure to find something even for demanding children.

Babies and young children don't care how you dress them, and you can choose from a wide range of costumes. The most popular are beetles and flowers. Get cute little one-piece dresses that form the body of the beetle-like a ladybug, a firefly, a honeybee, an ant or even a butterfly or a snail. As an accessory, you also get an antenna for each of the beetles as a cap or as a hairband accessory. All accessories are well padded, so you can be sure that your kids are safe and that they cannot get hurt.

There are similar. A range of flowers to choose from. A large part of the costume consists of the headdress, with the body forming the stem and leaves. You will find several comfortable clothes for your children to slide. Remember that you are looking for one that has zippered openings in the right places to allow for easy diaper changes. Kids are also wonderful models for food-based clothing, and you can find costumes like Hershey's chocolate or even strawberries, bananas and apples to choose from.

When it comes to children, the range is striking. Of course, you can start your search with a wide range of Disney costumes. Choose any character from the Mickey Mouse Clubhouse or the Power Puff girls or Winnie the Pooh and the Friends of the 100 Acre Wood. Children can also be dressed in costumes, depending on what they want to be. Here you can see doctors, nurses, firefighters, astronauts, scientists, dancers, engineers and much more.

There are various other ranges of children's Halloween costumes. You could take a look at a wide range of pirate clothes or maybe Ninja Turtles. Girls can be dancers or nurses or teachers. They can also be fairies and famous people, just like adults.

HALLOWEEN COSTUMES FOR KIDS

The Halloween costume for kids is available in more shapes, sizes and colours than anyone can imagine. You can find Halloween costumes for babies and children in all variations, shapes and characters. If you want them to dress like little pandas or princesses, you will find the right costume for your little one.

The selection of options that you can see these days in the kids' Halloween online store is simply amazing. There are a variety of favourite costumes of all time, such as Barbie or Snow White or costumes that pretend to be princesses, pirates and other fairy tale characters.

There are also many Halloween costumes for cute kids, and the hardest thing parents need to do is decide if their baby would look prettier in a hornet costume, a small pony costume or a pirate captain child costume. Other popular Halloween costumes for kids are princess and fairy costumes such as the Tinker Bell dancer costume and Barbie costumes.

The latest theatrical releases and popular action figures are always favourites from children's Halloween costumes. For example, Anakin Skywalker, Darth Vader Costume and Padme Amidala Child Costume are for lovers of Star Wars films; cartoon shows inspire Scooby-Doooo, Power Rangers and the Tank Engine Costume. Then there is the Kid Halloween superhero costume. Kids love dressing up as their favourite superhero for Halloween and during recreation and on special occasions. Choose from characters from the movie like Superman, Supergirl, Batman, Spiderman, Mr Incredible Costumes, Harry Potter, the Harry Potter Hermione costume, Wonder Woman from the Justice League and the list goes on. These are some of the most popular Halloween costumes for children, chosen for their popularity and the attractiveness of fans among children.

Parents can also choose from the award-winning Halloween costume for kids like an astronaut costume, a Champion Racing Suit Jr. costume and a dragon costume.

If your baby is a girl, there are beautiful princesses, Ariel costumes, Elizabeth of the Pirates of the Caribbean, Little Mermaid, a medieval princess, a pirate girl, a pirate queen, a beautiful witch costume and a first dancer to choose from, Snow White, Glinda Magician of Oz Costumes with a clown suit!

If he is a boy, then the variety is the most action-packed variety/superhero with the knight of the dragon costume, the mummy costume, the ninja emperor, the motocross rider, the Shrek, a pirate captain, a pirate king and a tin man (from the wizard of) Oz), Warrior King, Wizard costume, The Fantastic Four thing, Robin, Zorro and much more!

However, Halloween is never complete without creepy costumes, and the most popular creepy costume ever is the vampire costume. There are other creepy Halloween costumes for kids like the Sailor of Death costume, the Gothic Pirate King costume. Then there are various accessories such as Dracula teeth or long nails that can be used as accessories with these costumes and allow children to scare their friends and even their older sisters and brothers!

Seeing kids' "tricks or tricks" in their Halloween costumes is the most popular part of Halloween, and it's worth it for them to navigate through the myriad of colourful options. Whatever your child's character choice is, check out oyacostumes.com - you'll have something for yourself!

CHOOSING THE BEST CHILDREN'S TELESCOPE - TIPS FOR BEGINNER TELESCOPE BUYERS

One of the best gifts you can give a child is his first telescope. Nothing beats the excitement of looking into space for the first time. The first glance at the moon and all its craters and the sight of the rings of Saturn surprise children and adults alike.

Choosing a telescope for a child is difficult. You want to make sure you choose quality. Toy stores offer a selection of telescopes, but the sad fact is that most of them are rubbish that barely works. A little research before buying can make a difference. There are high-quality, affordable and easy-to-use riflescopes for beginners. On average, you can expect to find retail quality ranging from $ 75 to $ 100 or slightly more. By researching and comparing prices on the Internet, you can get even better deals.

The most important part of a telescope is the objective. Most starter telescopes have a so-called "refraction lens". These lenses magnify objects using convex curved glass. They are adjustable for focus and field of view. In this way, the user can point the area on an object in the room and focus the area on the object.

If you buy a beginner telescope, you can also give a guide to the stars and planets. There are many good books for children and young adults that help them understand more about what they see through the realm. The first thing most people want to see with a telescope is the moon. After examining the terrestrial satellite and all its craters, the next stop is probably played by Saturn or even by the big car and the small car. Twins and Orion also offer an exciting view with a new backyard telescope.

There is no better way to promote love for learning the gift of practical experience. When you buy a beginner telescope, you are doing much more than just a gift. They release curiosity and interest in science that will stay with a child for years to come. Who knows, you might even inspire the next generation of astronauts!

SPACE THEMED BIRTHDAY PARTY IDEAS FOR KIDS

A birthday party in space is not difficult to organize. Depending on your budget, many great ideas can be implemented. The idea is to arouse children's interest and entertain them. Games with a space theme may also be included. Kids love moon races, space treasure hunts or treasure hunts. A competition to build your spaceship would also be interesting. If the party takes place at night, the environment can be further enhanced by placing glowing stars in the dark or bright posters of the universe and beyond.

In the part of the house where the party takes place, various planetary devices can be installed with which different elements of the room can be displayed—some of the more advanced planetary devices like a CD that allows children to visit the cosmos.

Another activity would be the creation of children of space stones. Give the children small pieces of stone and many washable colours so they can paint the stones in the colours they prefer. However, if you plan on having them painted, inform the guests' parents in advance if the parents want their children to wear a lab coat.

Space Trivia is a fun game for children. Remember to prepare the prizes for the team who can answer most questions and then explain the winner. In this way, entertainment is combined with learning. For creative kids, hand out paper plates, coloured pencils, markers and other craft supplies and encourage them to create chamber masks. Give the child the most impressive prize.

Science fiction and space facts can help us a lot in planning the party. Children can try the different types of space food available for astronauts. This is both educational and fun. Children can learn which foods are consumed in space and why astronauts cannot eat normal and traditional foods in space. Some foods like brownies and fruit can be consumed in their natural form. Other foods require the addition of water such as macaroni and cheese or spaghetti. An oven is provided in the space shuttle and space station to heat food to the correct temperature. Explain to them that there are no refrigerators in space. Therefore, food must be properly stored and prepared in space to avoid spoilage, especially during long-term use.

Spices such as ketchup, mustard and mayonnaise are available. Salt and pepper are only available in liquid form. This is because astronauts cannot spread salt and pepper on food in space. Salt and pepper would simply swim away and pose a hazard because they could clog the vents, contaminate the equipment or get stuck in an astronaut's eyes, mouth or nose. Drinks range from coffee, tea, orange juice, fruit punch and lemonade. By explaining these facts, children not only learn and enjoy your space party, but they also have stories to tell when they return home.

Kate Slinger is a professional event organizer and enthusiastic author. Use common sense to organize parties, celebrations and festive events with unique ideas and a limited budget.

DESCRIPTION OF THE ASTRONAUT JOB

The work of an astronaut is in itself stimulating and innovative. Everyone dreams of flying high in the air and being an astronaut to make this dream come true. Astronomy is a branch of science that deals with the flight of light aircraft and flight flights. You can fly high in the blue sky and go to places that most people can only think of. You can have fun escaping from Earth's gravity and move freely, especially enjoying scenes such as the Great Wall of China and other wonders.

Types of astronauts

Astronauts are highly skilled scientists or engineers who go beyond the Earth's atmosphere to explore the outside world and learn the secrets of nature. The field of astronomy is expanding by leaps and bounds, and there are no limits to its exploration. Astronauts use space shuttles or other technical devices to enter space.

There are three main workspaces for an astronaut. Pilot, mission specialist and payload specialist are three different areas in which astronauts can work. Working as an astronaut can be both physical and psychologically demanding and difficult. You have to travel long hours in stressful conditions that are usually unbearable for ordinary people. Astronauts receive rigorous training to adapt to changing environments outside the Earth.

Responsibilities of an Astronaut

An astronaut is responsible for the regulation and operation of all space stations of the space agency. He also managed rhythmic missions and trips to explore the outside world and new planets. It is a very difficult job, which is always associated with the risks of life, but at the same time, it is exciting and adventurous for those who have a passion for space and other heavenly bodies. Astronauts do great jobs and need to be highly concentrated and extremely bright to benefit from space missions and other astronaut activities. You should have at least 1000 hours of flight experience on the plane to catch a space shuttle. The job description for astronauts is very well defined for students who are looking for a career and can consult the Internet.

LET AN ASTRONAUT TELL YOUR CHILDREN A FAIRY TALE

Space travel has fascinated children for generations, from children watching the moon landing in 1969 to girls learning about Sally Ride in 1983.

This generation is no different. Instead of hearing stories of people in space from journalists here on Earth, this generation has the ability to communicate directly with astronauts as they take spacewalks and orbit our home planet.

Protect your helmets, fasten your seat belts and prepare your children for Story Time From Space.

What is Story Time From Space?
Story Time From Space is a progression of recordings made by space explorers on the International Space Station. Against the backdrop of the earth, astronauts read from a collection of children's books and teach children about space in a fun and creative way. Teachers and parents love to see their children's eyes light up and see a real astronaut share some of their favourite books.

Story Time From Space is an initiative of the Global Space Education Foundation, a non-profit educational institution that works for all areas of the space sector. Since 1983, he has been working to connect, educate and inspire people to be involved and support the global space community.

What books have been sent to space?

As you can imagine, most of the books read in Story Time in Space have a specific theme: science and the great beyond!

An author who has been highlighted on various occasions is Dr Jeffrey Bennett. Bennett sent six of his books to the space station and has been a supporter of the program from the earliest starting point. He made a progression of books with Max, a pooch and his companions who travel through space and know the universe. Up until this point, Max has gone to the moon, Mars, Jupiter and the space station, where astronaut Mike Hopkins perused his story.

Guardians and teachers who need to follow the Story Time in Space arrangement can check the site consistently to perceive what's coming. At the top of the list of the International Space Station is "The next time you see a sunset" by Emily Morgan. This is part of their "Next Time You See" series, which includes books on fireflies, ladybugs, shells and the moon. Books invite readers to learn something they see every day and don't think about.

"Next Time" is followed by Jennifer Rustgi's "A Moon of My Own", a book about a girl who travels the world and learns about nature with her faithful companion - the moon.

The books on Story Time in Space must not be directly connected to space, but face the curiosity of children to know the world around them.

The space shuttle flies over the earth.

How was the time of history recorded in space?

So far, kids can't get enough of the stories read by their favourite astronauts.

In an article for Edutopia, Ben Johnson says that the response to the program has been largely positive and that the organizers have received hundreds of emails worldwide. This inspired her to go further. Now astronauts are conducting scientific experiments with readings to further combine the worlds of space, reading and imagination.

The popularity of the program also sparked a conversation about the importance of storytime in the classroom and at home.

"The 'magic' lies in the way stories deal with children's hearts and minds as people and thinkers on relevant, real and important issues for them," says author Carol Reid. The room is nice, but the story content keeps the kids busy until the end.

Other education professionals agree. "When we read, we share experiences, hopes, joys and worries while immersed in a good book," wrote the Barbershop Books team. "When mom, dad, aunt and uncle express their surprise at the big turn in the plot of the book, the children notice it and are even more excited."

Astronauts can establish emotional connections with children on Earth by sharing a reading journey with them - connections with science and space that are not easy to solve.

The Andromeda galaxy

Is there any other storytime shows that will follow?

Fans of Story Time in Space may be looking for other shows to help their children discover a new favourite book or learn valuable lessons. Here are some of our favourite moderators and programs that you can watch to bring books to life.

Bedtime FM is an independent podcast producer with Story Time, a podcast with bedtime readings from guest authors. The podcast never lasts more than 20 minutes to keep younger students busy.

Storyline Online was developed by the SAG-AFTRA Foundation and offered actors who read children's books, accompanied by illustrations. Viola Davis, James Earl Jones and Betty White are just a few actors who participated in promoting literacy for the next generation.

Author Brad Meltzer has a series of StoryTime videos with readings and re-enactments of his "I Am" series on his website. Barbara Bush even went so far as to read "I Am Lucille Ball" to support her literacy foundation.

Younger students may appreciate Story Time With Ms Becky, a YouTube series for children ages 3 to 10. In the end, Becky Bear's assistant asks questions that help readers critically review what they have just learned and the lessons behind the books.

Sydney Soli developed Storytime Yoga for Kids, in which parents and children extend through yoga positions and tell multicultural stories.

Storytime has the power to introduce children to new worlds and ideas. It can arouse curiosity about the world and even space by introducing characters and sharing adventures. Discover Story Time From Space and the other resources above to stimulate your child's imagination and motivate him to learn.

UNCONVENTIONAL WISDOM FOR ASPIRING ASTRONAUTS

If you've managed to keep your dream of becoming an astronaut alive, let's take a look at some things you can or must do to get your dream job. It all started as soon as you sneaked into NASA.

The people in the top positions at NASA are a group of tough people who expect the best from their people. The first thing to do is to see if you are the best candidate for the job or not. Don't worry, even if you're not the best candidate for the job. It doesn't matter if your only experience is watching cartoons or starting a shuttle on TV. I'd like to share a secret with you here: if you've only seen one shuttle so far, you can move.

Pretend

The spacesuit is the first thing you need to become an astronaut. What you can do is visit your local costume shop and see what they have for you. You can purchase an extractor hood for work for just under $ 50. After putting on your suit, you're ready to go into space. Go to the NASA office and show them the fake security pass. You will think you are a perfectly qualified astronaut and let yourself in without knowing what hit them.

Discover the beginning

You will surely be worried if you sneak into the NASA business with no experience and you don't know what you will do. So leave your worries aside as starting a shuttle isn't that difficult. You will be overwhelmed by all the livers and the crazy jargon that spews you through the headset. You will find a small red button labelled "Start". You just have to push it in and out.

Space Travel

After a successful space trip, NASA can now do nothing to stop you from doing what you want. If you think you are good at pretending, you can tell them that you cannot start another astronaut because he is not qualified enough. It is a good idea to launch an additional crew member so that you have more space to collect some interesting things that you collect from space.

come home

So far, so good and now is the time to go home. Maybe you are wondering how to do it. Ask one of your fellow astronauts why there are so many astronauts at the beginning. Here you may have to reveal you are true identity. Are you getting angry? Not, everyone wants to go home, so it can't hurt to take you with you.

THE ASTRONAUTS WIVES CLUB - A TRUE STORY

People magazine wrote of The Astronaut Wives Club A true story: "The men who were catapulted into space in the 20th century were quite interesting. The women they had left on Earth were fascinating."

In book confirmations, author Lily Koppel writes: "I still find it surprising that my iPhone contains more computing power than the technology that brought astronauts to the moon."

Many of the future female astronauts had been military women during the war, and they were familiar with long missions. Becoming an astronaut woman did not change her husband's long absence from home but brought them the power of the stars through association. The women who lived behind the photos and articles in Life magazine found their lives full of stress, glamour and benefits that did not simplify their lives.

Of course, the call to pretend to be the "perfect family" that NASA had imposed on them had an impact on their marriages. Most couples divorced after the astronaut went into space. It was pressure for her children to have a part-time father who was a public hero. They barely knew the children who rarely saw their fathers.

Before being chosen as astronauts, the elect were regular military officers whose wives transformed lives around husbands and children. They moved from the military base to the military base as required by her husband's career. Like astronauts, the men were military men who maintained their rank but were loaned to the "new civilian space agency", and the men no longer wore military uniforms.

Astronauts travelled frequently and lived in many time zones. When they were at home, many turned off their feelings and believed that a good driver was emotionless. A good pilot and a good astronaut must be alert and ready for anything because they must be able to make immediate decisions.

"... the women experienced the same awake nightmare and imagined the dark shape of the basement chaplain who rang the bell and told her that she was now a widow." The astronaut lived daily and knew that if one of his husbands had lost his life on duty, he would have had to calmly and courageously deal with his death. It was part of their job. NASA ordered it and the world also looked - judged.

The Astronaut Wives Club isn't just the story of exceptional women who came to the fore when their husbands became the first American men in space. The plot is a time capsule of family life in the 60s and women's libraries of the 70s. Two decades in which American families have changed dramatically.

When Lady Bird and Lyndon Johnson arrived at the home of astronaut John Glenn and his wife Annie for dinner, Annie served her humble but popular Ham Loaf. Annie has been highly praised by other women who have seen her as the friendly type of Betty Crocker. Before Lady Bird left Glenn's house, she thanked Annie and asked for her ham loaf recipe, as Annie knew she would.

The first female astronaut, Marge Slayton, who died in 1989, once told the Astronaut Wives Club that it was founded: "We all found our way through an experience that was the first - that of being female astronauts and feeling home with husbands we have to support each other. So we started with monthly coffees. "

The women who make up the Astronaut Wives Club have the "right things". This book about her describes her story well. A story interesting enough to become a popular 2015 TV series called The Astronaut Wives Club. Book or TV series This is a historical time capsule that everyone can learn from, especially those who want to take a look at this era.

HOW TO GET BIGGER BY IMITATING ASTRONAUTS

After a few days in space, astronauts begin to experience an extraordinary phenomenon: they become larger. In fact, what is happening cannot be exactly described as "growing", but the bottom line is that they are bigger.

The reason for this interesting and coveted salon trick is the lack of gravity they experience. In a nutshell, there is no gravity in space to "break down" astronauts, relieving their thorns from the weight of their bodies.

While this phenomenon was expected, the magnitude of the neglect shown by the astronaut's thorns shocked even NASA's most stubborn doctors when astronauts were examined in the early NASA years after the flight. Frankly, they shouldn't be as big as they are. Some growth was expected, but certainly not the three inches or more it showed.

However, the effects were temporary when astronauts returned to Earth and maternally embraced Earth's gravity. It only took a week or two to return to their pre-flight height.

However, physiologists were inspired by the sheer size of the unexpected lengthening of the spine and soon studied the ability of the human spine to adapt in strange environments.

The human body is extremely adaptable, which means that concentrated, but safe, techniques have been demonstrated not only to allow people on Earth to experience similar but persisting effects.

The most important aspect to remember during the whole process is to always be safe with every exercise or stretch and never bend or stretch the spine in an unnatural position.

Your attitude also comes into play. While most people's posture is fine, many people may appear up to an inch or even shorter than they are due to an unnaturally sharp forward curve of the upper spine.

Proper nutrition is also of crucial importance since the lengthening of the spine is not maintained permanently, since the remaining space between the vertebrae is not filled by the support discs that separate them. However, these spongy discs can be strengthened and strengthened with the help of a generous daily supplement with calcium and other joint support elements. These include chondroitin, glucosamine and MSM.

RISKS FOR ASTRONAUTS

Everyone knows that when buying life insurance, risky jobs like acting, working on an oil rig or flying an aeroplane will significantly increase the price. What does this mean for astronauts: those heroes of us who choose to be blown up in seconds? Let's take a look at the risks.

I assume that we must first consider the space element. Humans are physically limited by the number of G forces they can withstand, and it would be possible (but very slow) to reach space while traversing only one G. Therefore, it is common that when rising between five and seven G, the target is to reach space in just eight minutes. No wonder one of the biggest deaths in space was to be mourned when the Challenger shuttle was launched in 1986.

Interestingly, they move just as quickly in orbit. But the risks to astronauts get worse when you consider the amount of radiation in space. Cosmic radiation can not only affect the blood marrow, but it can also weaken the immune system and increase the risk of infection, but it can also affect the drugs taken with it. According to recent reports, NASA is currently investigating the damage that radiation can cause onboard pharmaceutical products. Testers have found "significant degradation" in some drugs, including Augmentin and Bactrim.

The effects of weightlessness are also not without negative effects, although they appear to be one of the most attractive factors in space travel. Tests have shown that astronauts who have experienced zero gravity for an extended period are known to develop loss of bone density, lack of muscle strength and endurance, postural instability and aerobic capacity. It is believed that an astronaut who would have ever travelled to Mars would have simply collapsed on arrival due to his absence of gravity.

If space travel is so dangerous, how much does life insurance cost for astronauts? Insurance specialist Ed Hinerman says that a person who becomes an astronaut will need to be reevaluated and will most likely have to pay a lump sum for every £ 1,000 of their policy. He claims that if I were to pay £ 2,000 a year, an astronaut might have to pay a flat rate of over £ 16,000 a year. It must be worth it.

LIVE YOUR ASTRONAUT DREAMS AT THE KENNEDY SPACE CENTER

If you've ever dreamed of becoming an astronaut, you can make that dream come true by visiting the Kennedy Space Center in central Florida.

A short drive from Orlando, the Kennedy Space Center Visitor Complex offers a unique opportunity to visit launch areas, meet an experienced astronaut, see missiles, train in a space flight simulator or even see a missile launch or of a shuttle, It's all part of the fantastic adventure that awaits you.

The shuttle launch experience offers an exciting journey where you can feel what it's like to go into space. Fasten your seat belts and experience the panoramas, sounds and sensations of a real space shuttle launch.

After entering the heart of the shuttle's operations for a pre-launch briefing, you will be guided by experienced space shuttle commander Charlie Bolden, who will guide you through the shuttle launch sequence step by step. Brilliant video screens come to life on shuttle-like robotic arms. Anxious moments arise when atmospheric noises and light effects dramatize the moments before the start of the space shuttle. So wait and enjoy the trip! You can also travel back in time and history with a visit to NASA Rocket Garden. Dramatic lighting brings NASA's historic rockets to life, including the Redstone, Atlas and Titan rockets, which launched NASA astronauts into space for the first time. You can board the Mercury, Gemini and Apollo capsules - and get an idea of the cramped conditions that the pioneers of the American astronaut have experienced.

Then go back to today's space-age and get on the International Space Station. This fascinating attraction gives you a close look at the actual structure where NASA prepares the actual components of the International Space Station - the largest and most complex structure ever put into orbit.

In an elevated observation room, it is possible to see the current processing compartment, in which each component of the space station is extracted, processed and prepared for its journey into orbit.

After the observation room, you can enter a complete model of the housing module and see how the space station crew lives, sleeps and works.

In Space Shuttle Plaza you will find a full-size NASA Space Shuttle replica - Explorer. See for yourself how astronauts live and work aboard space shuttles. You'll also find other components needed to launch NASA space shuttles, such as a huge external tank and two solid rockets.

Next to the Space Shuttle Explorer is the Launch Status Center, where visitors receive real-time information on NASA's ongoing launch and spaceflight activities.

Even more obscure: the Space Mirror Memorial, designated a national monument by Congress and President George Bush and inaugurated in 1991, honours the 24 American astronauts who gave their lives to explore space.

The names of the fallen astronauts from Space Shuttle Columbia, the Space Shuttle Challenger and Apollo 1, as well as the astronauts from training and scheduled air accidents, are engraved on the 42-1 / 2-foot-by-50- of the monument. Black granite surface as wide as afoot, as if it were projected towards the sky.

DISCOVER DISTANT PLANETS WITH YOUR ASTRONAUT COSTUME

On August 26, 2010, NASA held a press conference to announce that the Kepler space mission had discovered an Earth-sized planet, along with two giant planets orbiting a star. Astronomers also discovered a planetary system with at least five, perhaps seven planets in orbit around a sun-like star. This means that the ability to explore distant worlds outside our solar system will be more likely in the future. An astronaut costume to bring people to these distant worlds should be something very special and designed for very long journeys.

Many scientists believe that in the future it will be necessary to find these distant worlds and eventually travel, as our planet Earth may no longer be able to support us. A mission to Mars would most likely be the first step in the process. It's easy to imagine being on such a space flight in an astronaut costume that should take you to the red planet and back. You may also see yourself in a new astronaut costume designed to stay on

Mars for a long time as we start living there and building a new civilization.

You can pretend to be one of the first people chosen to travel to distant worlds with an astronaut costume. Be the first person to enter an Earth-like planet in a newly discovered solar system in our Milky Way or a distant galaxy.

On a planet much closer to Earth, the moon, scientists and astronomers have recently made discoveries on the side of the moon that we can only see through satellites in orbit that make a mission to the moon possible. You can easily imagine yourself in a flight to the moon in your astronaut costume.

In March 2009, Russian scientists began a 520-day simulated flight to Mars by inserting a team of 6 researchers into a series of capsules that are expected to look like a spaceship on a long journey. President Obama said that astronauts could orbit Mars in the mid-1930s.

Today's children would be the first to see Mars up close. A Halloween astronaut costume could be a stepping stone for a royal outfit, as the first humans will enter Mars in the not too distant future.

China has also announced that it will launch the first part of its space station in 2011 and a possible manned mission to the moon in 2017. They also announced that they would select mothers to train the female astronaut, the jet fighter pilots in the Chinese air are powerful.

Attention is increasingly focused on space travel. An astronaut costume for Halloween reminds us that today's young people will be tomorrow's space travellers.

THE ROBOTIC ASTRONAUT ASSISTANTS

Currently, the only operating robotic assistant astronauts are the remote manipulators of the Space Shuttle and the International Space Station. These remote-controlled robots are used as crane-like manipulators to transmit payloads and EVA astronauts. The focus of the coming decades in the exploration of human space will be on the surfaces of Mars and the moon. This means that new types of astronaut assistants are required, particularly on Mars, where teleworking from Earth is not possible. The first step in this development is to evaluate the new requirements for the corresponding technology and the activities of the robot assistants. They can be used to develop robot demonstrations to test the usefulness of identified technologies in practice.

What is an assistant robot astronaut? An assistant is a person who helps meet a need or promote an effort or purpose. A robot astronaut assistant is a robot actor who contributes to making an astronaut effort. This relatively free definition is sufficient to trigger important follow-up questions. What advantages could the assistant offer, what kind of robot could be used for support?

Cooperation, collaboration and coordination are related terms that are often used to describe ambiguously how humans and robots carry out activities together. Instead, the work was divided into independently solvable sub-activities, and coordination between the actors is only necessary to combine the results. Coordination can be defined as a simplification such as managing dependencies between activities. This means that the collaborative and characteristic actors elaborate the joint work in the next course, while the cooperative actors focus on the correct execution of the defined joint work. Astronauts and robots work together to carry out activities that include tasks, missions and hierarchical actions.

The potential of robot astronaut assistants has been recognized. For the time being, both NASA and ESA have identified critical roles for various types of automated and robotic technologies in their future space exploration missions.

ASTRONAUT COSTUMES ARE A FUN WAY TO PAY HOMAGE TO OUR HALLOWEEN STORY

The astronauts' costumes have long been fashionable on Halloween, and young people dream of maturing in space travellers. Due to extraordinary astronauts like Barbara Morgan, more and more young women are enjoying themselves in disguise as children's astronaut costumes and dream of being next to walk on the moon or Mars. Astronaut costumes usually enjoy great popularity whenever NASA companies are in the media and space movies are always in trend. Astronaut costumes for kids will always be a hit, so kids can use their creative imagination to pretend to explore areas of space never seen before or rebuild the famous moon landing.

Neil Armstrong's costumes will all be popular. Neil Armstrong, the NASA astronaut, whose famous phrase has been heard by practically every child in the United States since it was uttered on July 21, 1969, when he first landed on our celestial satellite, better known as the moon. Armstrong's moonwalk has encouraged kids around the world to choose astronaut costumes for their Halloween fun year after year. Given that his famous phrase was aired on stations around the world, it can be assumed that over 450,000,000 people listened to his statement, which automatically generated a group of aspiring astronauts. Armstrong, like most previous astronauts, began to become an army test pilot. The previous astronaut suits that NASA men needed were almost nothing compared to the modern space suits available for today's astronauts. Today they are much lighter, much more sophisticated and easier to wear for longer periods. Of course, the space shuttle costumes are certainly not as sophisticated for people stuck here on Earth, but it's fun to imagine that you will sometimes join Mr Armstrong on his Apollo 11 mission.

Various popular space movies further stimulated the attractiveness of astronaut costumes. Popular space cartoons have become increasingly popular among children. At the same time, adult films like Apollo 13 and Armageddon have also made adults wear astronaut costumes and believe they are on a mission to save the world from some destruction. As a result, the spacesuit costumes offer children and adults the opportunity to pay homage to Neil Armstrong and others who for many of us represent the ideal of space travel.

Space shuttle costumes tend to be a fantastic choice for theme parties. This is especially true of children's birthday parties. If you wear all your little participants in astronaut costumes, the photo opportunities will be exceptional.

DISCOVER NEW THINGS IN AEROSPACE TECHNOLOGY

The field of aerospace technology is very broad. It includes a wide range of disciplines in the aerospace and aeronautical technology sectors with applications for the various specifications, designs and constructions of engineers of aircraft, satellites and other spacecraft.

If a person wants to become an astronaut, one must first determine whether this professional path is suitable for you. If the descriptions sound as good as you, you are probably suitable for a career as an astronaut.

Astronautics encompasses the dynamics and other areas of advanced science and space technology. Space is becoming increasingly important for our economy, national security and research. Space engineers design and build missiles and missiles, space launches, communication and direct satellites, space navigation systems, remote sensing, spacecraft for human space flight and planetary probes. They also operate a ground circulation system from the ground control system.

The technology is intellectually suitable and combines the basics of science and technology with special astronautic knowledge. Graduates from these companies are ready to join the space industry and state-space research and development centres.

To obtain an astronautic apprenticeship, an engineering degree or a mechanical engineering degree with a career option focused in an aerospace engineer is usually required.

Engineers who want to specialize in aerospace engineering should have a solid foundation in math and science in high school. Students should also have a scientific curiosity to be both enterprising and analytical in their thought process.

Tasks of the astronaut

• Participate in participation in research and development programs

• Prepare a technical report and other documentation such as manuals or bulletins for technical personnel

• Keep detailed records of equipment, materials and performance

• Evaluate and modify existing projects by formulating mathematical models or using calculated methods

• Planning and execution of environmental, experimental or operational tests on existing models of aerospace systems.

Astronautical engineers are hired by organizations involved in the development and manufacture of spacecraft and related devices and components such as satellites, weapons and defence systems. They are also hired by aeronautical and space agencies, as well as executive agencies that focus on scientific research and engineering projects.

Usually, they must be approved as professional engineers because their work affects public safety. To be approved as a professional engineer, the following requirements must be met

• Completion of an accredited engineering program
• Evaluation of the basics of engineering
• Relevant work experience

Similarities between aviation and astronautics

Both technical options examine aerodynamics, structural aspects, propulsion concepts, as well as navigation and flight control. The main difference, however, comes from the main focus of the individual areas of aircraft and spacecraft. For example, the plane uses the air in which it flies to create elevators. Still, missiles primarily target the air as resistance and at much higher speeds and for a relatively shorter period of "missile time".

"ESCAPE FROM PLANET EARTH": FUN FOR CHILDREN AND ADULTS

Family comedies often focus so much on children's entertainment that movies forget adult viewers. "Escape from Planet Earth" offers enough laughter for children and their parents to enjoy the movie.

Scorch Supernova (Brendan Fraser, "The Mummy") is an astronaut who is loved by people of all ages. Whenever an alien planet kidnaps a person, Supernova hurries to help. He is helped by his brother Gary (Rob Corddry, "Hot Tub Time Machine"), who remains close to home with Mission Control. When an emergency signal arrives from the Dark Planet, Supernova says she should hurry up and help immediately, while Gary says his brother should be left behind and let someone else go.

Supernova contradicts his brother's wishes and runs to the dark planet. There, viewers learn that the dark planet is Earth. When an evil soldier decides to take Supernova hostage, Gary discovers that he is the only one who can save his brother and defeat the evil lord of the dark planet.

Children's films tend to retain their bright content, but "Escape from Planet Earth" actually has darker moments. A high-speed spaceship crushes a man and some scenes are cannibal extraterrestrials. After reaching the dark planet, Supernova meets a general who often makes fun of and tortures aliens who have found their way on Earth. These scenes maybe a little too dark for younger children, but prevent this film from being only for children.

The humour in the film falls on Fraser and Corddry's shoulders, and they do a great job of keeping the laughter going. Corddry plays the heterosexual man excellently, while Fraser excellently plays the action hero. Fraser draws from his work on films such as "The Mummy" and "Journey to the Center of the Earth", and while speaking a character who does not look like him, adult viewers will present him whenever the character is his. The mouth opens.

Adult spectators will also receive a kick from General Shanker, who heads the Dark Planet. The general voiced by William Shatner of Star Trek is crazy and funny. While the script requires a dark and sinister general, some viewers may recall Shatner's work on the Priceline.com commercials and smile when he starts talking. Shatner brings the right amount of humour, and the general becomes one of the most memorable characters in the film.

Some viewers may feel discouraged by the stereotypes represented in the film. Almost every female character is a sweet and loving woman who stays at home to raise children and cook dinner for her husband. Furthermore, scientists do not stray easily. Even Gary, who is the star and heart of the film, is unfortunate that others tease and call a nerd because he likes science. Her son is treated the same way, and some may find these scenes a little difficult to see.

"Escape from Planet Earth" is a kind of film in which the audience must listen carefully when the characters speak since the cast contains many high-quality actors. Ricky Gervais ("Ghost Town") appears in the role of Mr Bing while Jessica Alba ("The Fantastic Four") plays Lena. Viewers will also hear voices from Sarah Jessica

Parker ("Sex and the City") and Sofia Vergara ("Modern Family") in the film. These actors and actresses make you laugh and play similar and at the same time completely different characters from the other roles on the screen.

"Escape from Planet Earth" is a sweet story in the heart of two brothers. Almost everyone who watches the film will think of their families, especially their siblings. Gary is the brother who is so afraid of the world outside of Mission Control that he does not know what to do with his life and is even struggling to connect with his son. Supernova is so committed to saving others that she doesn't notice that she has problems at home. When Gary finally gets on the plate and leaves to save his brother, some viewers may want to leave the theatre and call a loved one.

The comedy in "Escape from Planet Earth" is suitable for all audiences. It's the kind of movie that parents can watch with their kids without fear of the boredom of childish humour, and they don't have to worry about their young children encountering adult jokes and situations. "Escape from Planet Earth" is a heartwarming story with lots of laughter for the whole family.

CHILDREN INTERESTED IN ASTRONOMY - SPACE GAMES FOR CHILDREN

How are children interested in something? Most children are curious by nature and if given opportunities and resources, they will be happy to learn and research, while others will need a little more encouragement and support. Astronomy and space fascinate many children - many of them will dream of becoming an astronaut at some point when they grow up.

How do you spark this spark of interest? If your budget allows, you can get them a $ 6,000 telescope! More realistic it could be a lot of fun using star maps and bright stickers in the dark to recreate a night sky on the bedroom ceiling. You could take them to a planetarium, and then there are small planetary toys that you can buy to have in your room (also make big night lights).

The variety of toys available for educational discovery science is enormous. The hard part will be deciding what to choose. There are stars, planets and stickers that shine in the dark or obviously. You can even let the entire solar system glow in the dark like a mobile or desktop display (motorized or not): the choice is yours. Uncle Milton makes a fabulous selection of space games that are also educational, such as "Moon in my room" and "Night Sky Navigator". Planetariums or star theatres of different sizes and produced by different manufacturers. So look around and get the best deals.

There are countless board games to choose from: they all introduce players to astronomy and space and allow them to have fun with their friends. I even came across a Monopoly Astronomy Edition. For the most serious young astronomer, how about a scientific exploration kit of the Thames and Kosmos, The Magic School Bus or the Young Scientist series. Kits such as those made by these manufacturers are excellent one-stop shops with educational brochures and everything you need to learn right away.

If you are considering a telescope, you don't have to spend $ 6,000! Children's versions start at $ 35 for an entry-level telescope. When buying a telescope, also consider a star map and a travel guide - there is also a large selection here. Take a look around and read customer reviews because they give you a good idea of what to expect before parting with the hard-earned money at Christmas or birthday.

Do you need more ideas? How about playing cards on the theme of space, an astronomy globe, a world map, puzzles, quizzes, a constellation board game, placemats, posters, inflatable solar systems and space shuttles.

A space encyclopedia is important reading for all young astronomers, astronauts or space fanatics

Dorling Kindersley is world-famous for publishing the best reference books, and its Space Encyclopaedia for Kids is one of the best. It must be the first perfect space book for children with the right amount of detail. With DK reference books, you always get interesting and high-quality information that is presented attractively and colourfully. On the 128 pages, the text is shown in different fonts and sizes. There are also test questions along the base of the pages with wrong answers on the opposite side, e.g. Why is there no blue sky on the moon? Because the moon has no atmosphere. o What does the word "comet" mean? Long-haired star. With the Curiosity Quiz, you can search for answers in each section.

The images offer an excellent balance between text, diagrams, photos and computer-generated images. The book has special features that show you how to get your hands on as much information as possible! Use the "Become an expert" button to learn more about a topic on other pages. The First Steps activity buttons show you how to try things on your own.

What is space?

Where does it start? Stargazers, observers, radio telescopes, Our space, Large galaxies, The Milky Way, nearby stars, The universe

Explore space

Astronaut in training, rockets, travel on the moon, men on the moon, space shuttle and stations, live and work in space, artificial satellites, explore Mars, reach the stars!

The solar system

Sun, eclipse, mercury, the third rock of the sun, moon, red planet, king of planets, the moon of Jupiter, Saturn, distant twins, Pluto.

Comets and meteors

Shooting stars, the asteroid belt, asteroid landing, space debris.

The secrets of space

UFO, is there anyone? Is there life on Mars? The Big Bang, black holes, are there other lands? A star is born, the death of a star.

Space for everyone

Become a stargazer, moon phases, constellations, northern and southern skies, space technology, spatial timeline

The universe was founded billions of years ago and is so large that you can hardly imagine how large it is. With this space encyclopedia, you can see how it was born, what is in the future and be amazed at the existence of alien life forms. Did you know that it hasn't rained on Mars for three billion years?

I can only recommend this book to a family who teaches at home or only to any parent with children who want to learn more about our solar system, our galaxy and beyond. The publisher's recommended age is between 9 and 12 years old, but children from 5 years old will be fascinated and happy to explore the book with an adult. It is full of information that curious young people want to know, and that is presented in an educational and fun way. There is a wide range of DK encyclopedias on subjects of the human body, nature, dinosaurs, science and animals, as well as an atlas, a dictionary and a general encyclopedia. All school libraries should have a range of these high-quality DK reference titles.

CONCLUSION

To paraphrase Oscar Wilde: we all live in the eaves, but some of us are looking for the stars and others more than astronauts flying in space.

The astronaut costume is a very popular theme for parties and Halloween, and a spacesuit costume is what many people look for online.

Maybe they have their dream of imitating heroes like Armstrong, Aldrin and Collins, the three guys who won the space race for the United States.

Undoubtedly, these three brave astronauts had looked up to the sky and had dreamed of really going there.

Today there are many different festivals where costumes, themed weddings, stag parties, Halloween and many more are worn. You could also have a Christmas themed package with packages that were distributed by an alien and received by everyone in disguise as a spaceman.

Children especially love to dress and a birthday party where everyone is wearing a costume is a great way to celebrate.

The problem with kids these days is that some cardboard boxes sprayed with glitter paint are no longer enough. Thanks to Hollywood and endless video games, but the truth is that children want realism.

An astronaut costume is easy enough to fake with some white coveralls and some NASA patches that mom needs to sew on, but accessories like the astronaut helmet require a little more investment. Do not cut a crack in an old paint can and spray silver.

To look good, the little astronaut needs some basic add-ons like a realistic helmet, maybe a NASA-style backpack, gloves and some authentic-looking space boots.

www.ingramcontent.com/pod-product-compliance
Lightning Source LLC
Chambersburg PA
CBHW071146240526
45465CB00024BA/1788